FIFE & TAYSIDE

A LANDSCAPE FASHIONED BY GEOLOGY

ISBN 1 85397 110 3

A CIP record is held at the British Library
L4K0201

Acknowledgements
Authors: Mike Browne (BGS), Alan McKirdy (SNH) and David McAdam (BGS).
Series Editor: Alan McKirdy (SNH)
Design and production: SNH Design and Publications

Photography
M.A.E. Browne/BGS 23, **A. Christie/BGS** 14, 16, **L. Gill** 6, 8 left, 8 right, 9, 10, 12, 13 bottom, 15, 17, 19 bottom, 20, 21, 22 right, **Natural History Museum** 13 top, **By kind permission of the Trustees of the National Museums of Scotland** 11 top, 11 bottom, **P&A. Macdonald** front cover, frontspiece, 4, 7 top, 7 bottom, 18, 22 left, **D.C. Thomson Ltd** 19 top, **Dr D.G. Woodhall** 5, back cover

Illustrations
Iain McIntosh contents, **Craig Ellery** 2, 3.

Further copies of this booklet and other publications can be obtained from:
Scottish Natural Heritage,
Design and Publications,
Battleby, Redgorton, Perth PH1 3EW
Tel: 01738 444177 Fax: 01738 827411
E-mail: pubs@redgore.demon.co.uk
Web site: http://www.snh.org.uk

Cover image:
Old Course at St Andrews showing stabilised sand dunes and the West Sands. This most famous of landscapes developed over many thousands of years. It progressively built out from the rocky coast as tidal currents dumped layer upon layer of sand along the foreshore to be reworked by the wind into sand dunes

Back page image:
Curved columns of basalt at Orrock Quarry

FIFE & TAYSIDE

A Landscape Fashioned by Geology

by

Mike Browne, Alan McKirdy and David McAdam

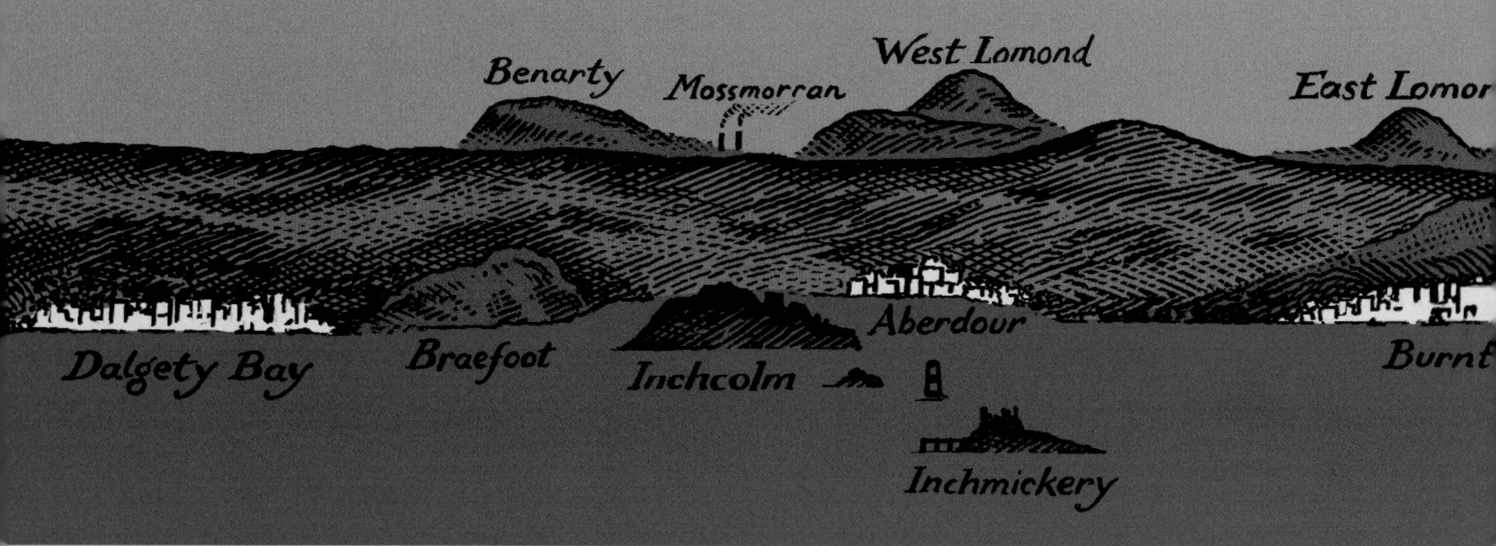

The South Coast of Fife and beyond - looking northwards

Contents

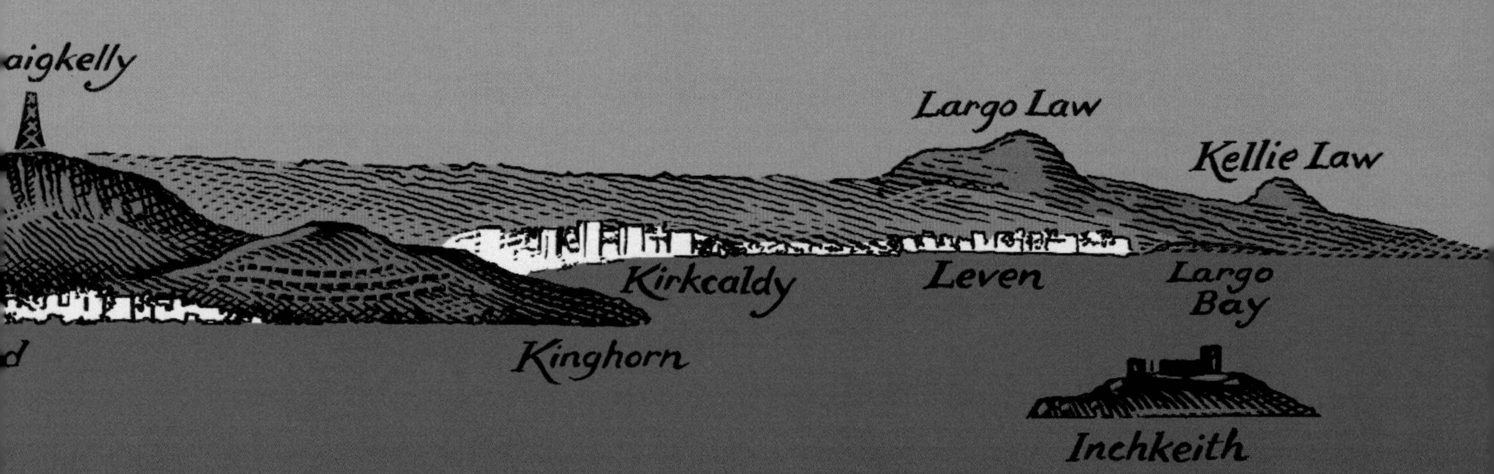

The gently undulating landscapes of Fife and Tayside mask a turbulent geological past. Rocks, fossils and landforms of the area tell of baking deserts, tropical rainforests and erupting volcanoes. In the more recent geological past, the area was covered by ice which shaped the contemporary landscape. This process of change and renewal continues to this day, driven by Man and nature. Along the coast, the relentless pounding of the waves moulds the coastline into rocky headlands and sandy beaches, whilst on land, we have modified the landscape for our benefit by draining wetlands and creating opencast coal mines. This book explains the geological processes that formed this landscape and the enormous timescales over which this part of Scotland was fashioned.

Fife & Tayside Through Time

Period	Events
RECENT TIMES	**250 years.** Industrial Revolution; land improvements with removal of peat and reclamation of lochs; burgeoning of mining and quarrying. **600 years.** Coal mining starts. **2,000 years.** Roman occupation.
QUATERNARY PERIOD THE "ICE AGE" **2.4 million years ago up to and including recent times** 	**5,000 to 4,000 years.** Sea levels fall to those comparable with present day; Neolithic hunters kitchen midden found; forests cleared for farming. **6,500 years.** Sea levels rise 10m above present day depositing layer of 'carse' clay around the coast, particularly the Tay Estuary. **9,000 to 8,000 years.** Peat accumulates as climate becomes warm and wet. **11,000 to 10,000 years.** Climate again arctic in character; freeze/thaw encourages landslides and blockfields to develop; sea level falls to present day levels or below. **13,000 years.** Ice retreats to Highlands; sea level is up to 45m higher than present day. **19,000 years.** Ice extends many miles east of the present coastline. **27,000 years.** Last advance of the "Ice Age" as ice builds up in the Highlands. **2.4 million years.** Climate cools and "Ice Age" begins; present day river systems established, later modified by glacial erosion.
TERTIARY PERIOD **65 to 2.4 million years ago** 	**62 million years.**Wideneing of the North Atlantic Ocean; no record of any geological events in Fife and Tayside.
CRETACEOUS PERIOD **65 to 135 million years ago**	**80 million years.** Warm shallow temperate seas fringe the land, with chalk deposited across Scotland, but later removed by erosion.
JURASSIC PERIOD **135 to 205 million years ago** 	**205 million years.** Opening of the North American Ocean begins. Climate on land is arid; desert conditions widespread.
TRIASSIC **205 to 250 million years ago** 	No record of any geological events in Fife and Tayside.
PERMIAN **250 to 290 million years** 	**290 million years.** Minor volcanic activity.
CARBONIFEROUS **290 to 360 million years** 	**295 million years.** Intrusion of Midland Valley Sill and associated dykes; subsequent movement on Ochil Fault. **305 million years.** Movements within the Earth's crust, causing folding, faulting, uplift of the crust and subsequent erosion. **325 million years.** Scotland sits astride the equator. Warm shallow seas fringe the land, reefs build up offshore. Tropical rain forestd grow on the coastal plains, building to form thick coal deposits. **350 million years.** Small volcanoes begin erupting on land and later under the sea. These vents tapped magma sources deep within the Earth's crust – "Elie rubies", pyrope garnets up to 4cm in size, are contained in rocks derived from the mantle, some 70km below the surface. Volcanic activity and earthquakes continue throughout the next 50 million years. **360 million years.** Tropical coastal plain was occasionally flooded by the sea. Salt deposits, especially gypsum and thin cementstone beds laid down.
DEVONIAN **360 to 410 million years** 	**370 million years.** Wide coastal plain established, fringed by desert, providing evidence of extensive areas of sand dunes. Rivers existed which teemed with primitive fish life, now fossilised. Significant breaks in sedimentary record at Arbroath – evidence of contemporary earth movement. **410 million years.** Large volcanoes existed producing lava and debris flows which built the Ochil and Sidlaw Hills. The climate was semi-arid with very large river systems flowing south-west across the area which deposited thick layers of sandstone and also conglomerate. Fossils of primitive plants, some up to 3m in height and fish found in associated lake deposits.
SILURIAN **410 to 440 million years** 	**420 million years.** The earth moves when Scotland collides with England as the Iapetus Ocean closes. Extensive folding, faulting, uplift of the crust and subsequent erosion ensues. The Highland Boundary Fault is created as a terrain boundary between the Highlands and Midland Valley.
ORDOVICIAN **440 to 510 million years** 	Great thickness of shales and sandstones accumulate in the Iapetus Ocean, which separated Laurentia (including the land area now called Scotland) from Avalonia (including the land now called England).
CAMBRIAN **510 to 550 million years** 	No record of any geological events in Fife and Tayside.

Geological Map of Fife & Tayside

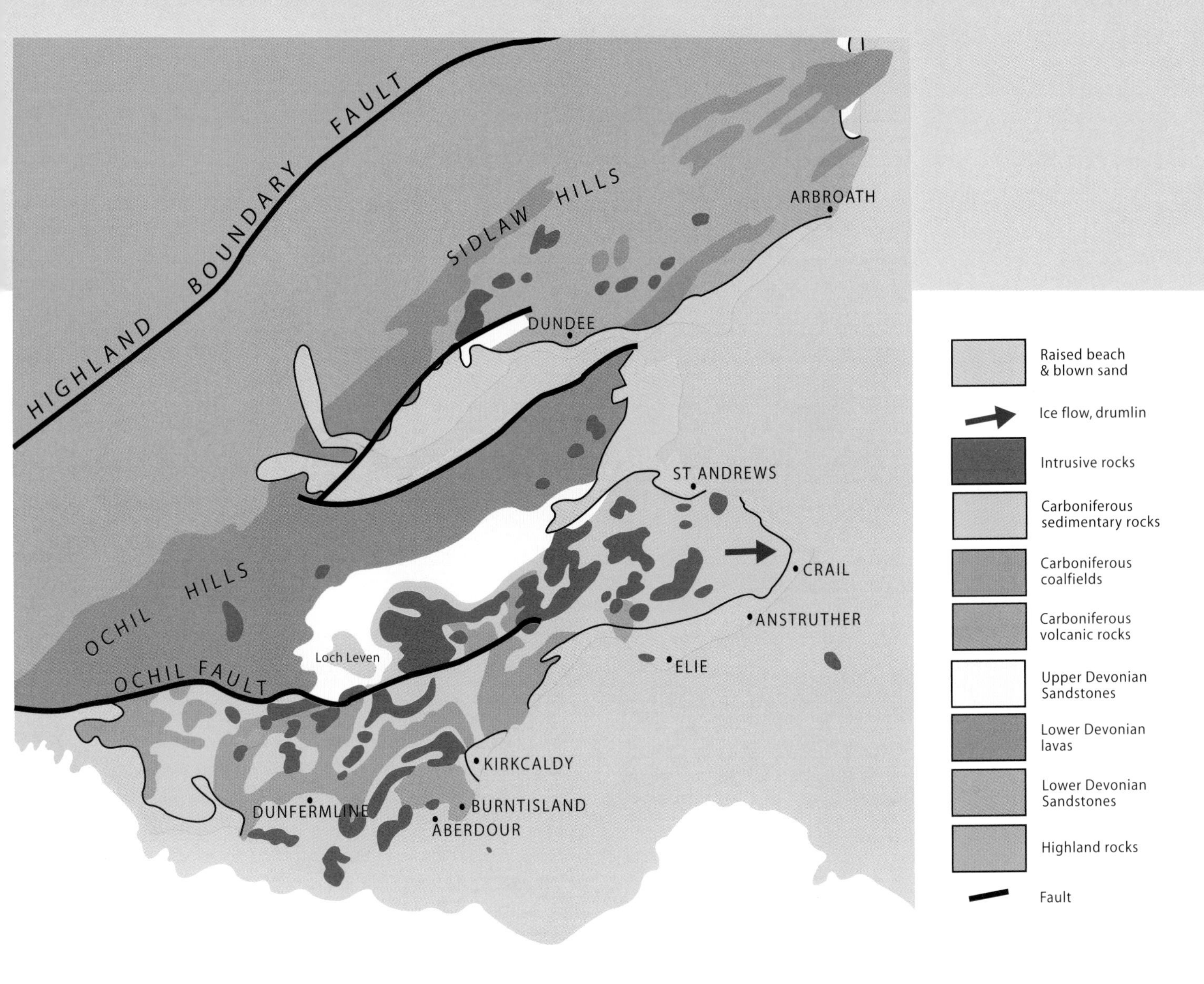

Kinnoull Hill - Lava escarpment

Volcanoes - Large and Small

Around 415 million years ago, Fife and Tayside were part of a hot, sparsely vegetated desert. To the north and south lay eroded mountainous areas. The Ochil, North Fife and Sidlaw Hills are what remains of the high ground to the south. They were once towering volcanoes that erupted magma - molten rock - from great depths in the Earth's crust. Some of the flows resisted weathering and erosion to form scarp features like those around Norman's Law and Newburgh in Fife, Craig Rossie, Moncreiffe and Kinnoull Hills in Perth and King's Seat in Angus.

Throughout Fife are the remains of smaller volcanoes active briefly between 350 and 300 million years ago. The hot magma and gases punched through a thick blanket of sediments laid down by meandering rivers, lakes and peat bogs. The resultant violent eruptions produced clouds of ash rather than lava flows. Where the eroded remains of the volcano is comprised entirely of volcanic ash, these deposits have made little impression on today's landscape. However, where magma cooled to leave a hard plug of rock, erosion has left conspicuous volcanic edifices like West and East Lomond, Saline and Knock Hills and Largo Law.

Where volcanoes erupted into the sea, as in Hawaii today, violent reaction between the hot magma and seawater produced large volumes of fine-grained volcanic breccia or fragmented rock. In more gentle interactions, the magma takes on a pillow-shaped form where the water chills the magma to a skin of volcanic glass. Pillows stack on top of each other to form spectacular formations, like those on the shore at Kinghorn. At Orrock Quarry near Burntisland, the submarine eruption created the most exquisite cooling joints as the magma froze beneath a layer of protecting breccia.

Orrock Quarry - columnar joints in basalt lava

Molten Rocks from the Deep

By no means all magma reaches the surface to form volcanoes. Much froze in the volcano's plumbing system below ground. Because the solid magma is much harder than the surrounding rocks, erosion has left many remnants as prominent features of our landscape.

At Ruby Bay, Elie pyrope garnets ('Elie Rubies') up to 4cm across occur in volcanic rocks which have tapped magma sources from deep within the Earth - to a depth of around 70km. These occurences offer a rare insight into the nature of the upper mantle of the Earth, as it existed during Carboniferous times.

However, some of the largest and most continuous sub-surface pulses of magma may never have been linked to any volcano. This is the case with the Lomond and Cleish Hills which are underlain by a complex of volcanic rocks, known as the Midland Valley Sill, extending across much of Fife.

The Rock and Spindle near St Andrews is the eroded remains of a small volcano

Lomond Hills - escarpment formed by the Midland Valley Sill and vents of East and West Lomond

The hard dolerite, which caps these hills is resistant to erosion, offering some protection to the easily eroded, underlying sandstones on the lower slopes. At Campsie Linn, north of Perth, the same type of rock is seen as a vertical feature intruded into a major tension crack in red mudstones. Such dykes represent routes along which magma flowed to feed the Midland Valley Sill. In places, the latter reaches a thickness in excess of 160m. Dundee Law is a much older feature related to the 415 million years old lavas in the Sidlaw Hills.

Dundee Law - an irregular sill of andesite

Desert Turns the Land Red

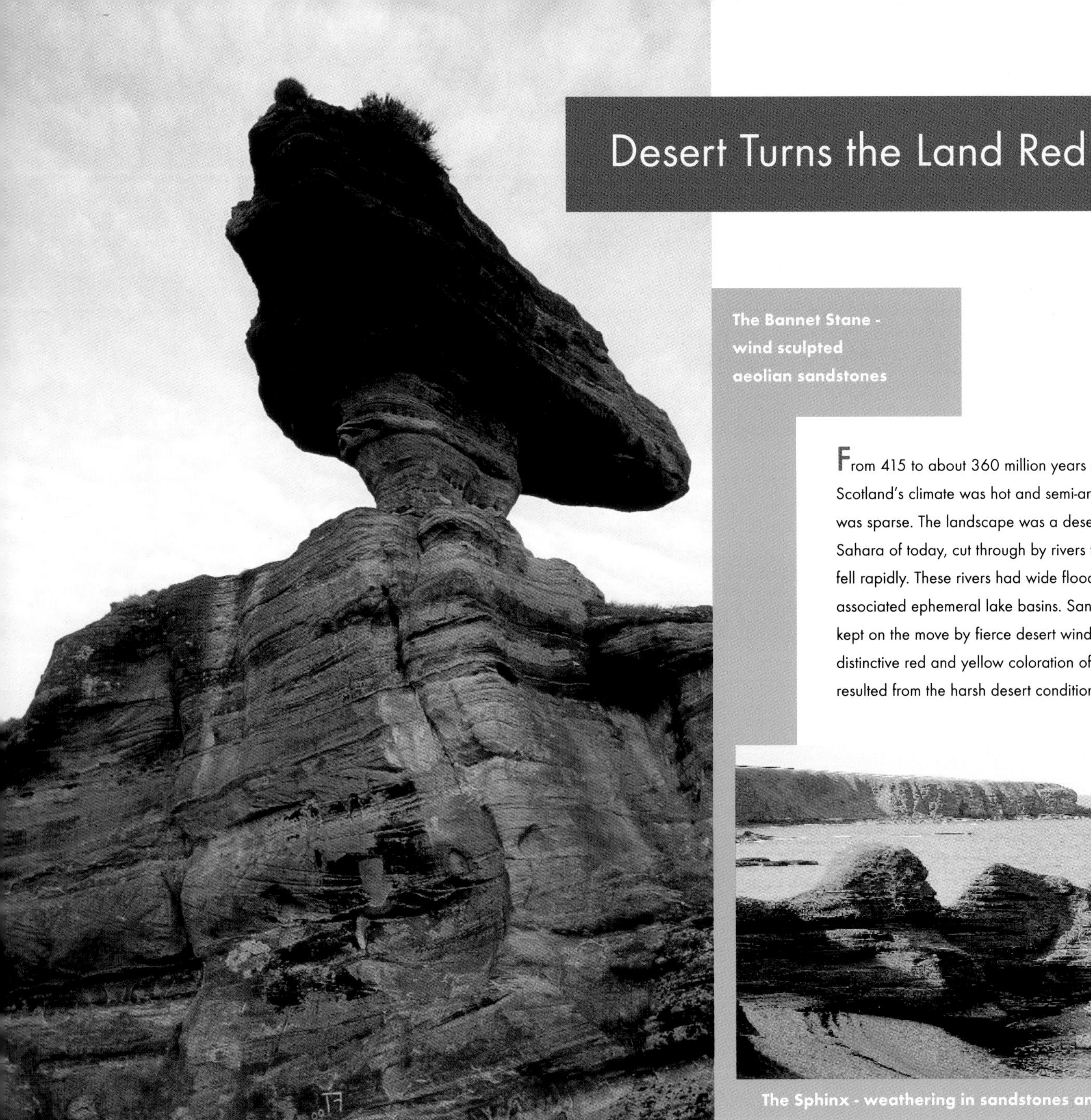

The Bannet Stane - wind sculpted aeolian sandstones

From 415 to about 360 million years ago, Scotland's climate was hot and semi-arid. Vegetation was sparse. The landscape was a desert like the Sahara of today, cut through by rivers that rose and fell rapidly. These rivers had wide floodplains and associated ephemeral lake basins. Sand dunes were kept on the move by fierce desert winds. The distinctive red and yellow coloration of these rocks resulted from the harsh desert conditions.

The Sphinx - weathering in sandstones and conglomera

Crossgates cutting - sandstones laid down by a river comparable in size to the Mississippi

The scale of ancient rivers that traversed the desert plains is indicated by the former river channel seen in the A9 Crossgates road cutting west of Perth. Here, the internal bedding structure of the sandstones shows that the river channel must have been in excess of 15m deep which would compare well with that of the modern Mississippi. The Crossgates River would have dwarfed the present day Tay and its catastrophic floods of 1990 and 1993. However, the braided island sections of the Tay above Perth at Scone would, vegetation aside, be a reasonable model for the ancient regime, albeit on a smaller scale.

Some of the oldest of these rocks originated as river shingle and related deposits. The conglomerates are particularly resistant to weathering and erosion, and form hills such as at Knock of Crieff, Craigowl north of Dundee and Hill of Finavon by Forfar. The Sphinx at Arbroath is composed of gravels and sands laid down in these ancient environments. The Bannet Stane, northeast of West Lomond Hill, and John Knox's Pulpit are composed of sand, blown by wind into ancient desert dunes. The sculpting of the Bannet Stane is partly the result of modern sand blasting by winds.

Tropical Limestone Paradise

The quarry face at Roscobie - mounds of limestone laid down in clear tropical waters with muds deposited by deltas on top. This site is in private ownership and access permission must be sought prior to any visit

From about 350 to 300 million years ago, Scotland lay near to the equator. Fife and Tayside formed part of an extensive, low-lying coastal plain which stretched far to the west and east. Environmental conditions changed dramatically over time. Sometimes the sea flooded the area, at other times there were extensive river and delta systems with lakes, freshwater swamps and coastal bogs resembling mangrove swamps. The climate was hot and humid. Lush vegetation in extensive rain forests consisted of rapidly growing trees up to 15m high buttressed only by a strong woody bark. Giant horsetails and club mosses, today represented by species of unimpressive stature, formed a prominent part of the flora.

In the warm seawater tropical shells, bryozoans, corals and sea lilies (crinoids) flourished. Though reefs were as common then as the tropical coral islands of today, reef-forming corals were a rarity. The reefs of the Carboniferous were built of microscopic algae, cyanobacteria and bacteria. These moundy build-ups were generally small structures, standing only a few metres and at most 15m above the seabed. At Roscobie Quarry, these mounds are overlain by muddy delta deposits which built seawards into the clear tropical waters.

The constantly changing environments of these Carboniferous times are reflected in the cycles of sedimentary strata laid down and preserved in the record of the rocks. The typical cycle begins with mudstone, laid down in estuarine to offshore seabed environments. Limestone is then deposited in clearer waters. Further mudstone is set down as deltas build out into the sea and the deposits coarsen upwards into sandstones. As the cycle continues, the sediment dumped offshore reclaims land from the sea and coastal plains are formed.

Hydreionocrinus anpuls - a sea lilly

Reconstruction of a limestone reef as it would have appeared during Carboniferous times

Coal Forests

This coastal environment was ideal for a tropical forest to become established and subsequently flourish. As the luxuriant vegetation died and decomposed, it formed thick deposits of peat. Although comparable to modern blanket and raised peat bogs which may be up to 6m thick, the Carboniferous deposits were up to 22m. These layers of peat were compressed by burial beneath other sediments to form coal, in seams between one to three metres thick. Because these sedimentary rocks are usually so much softer than the surrounding volcanic rocks, they form the low ground of Fife and the Carse of Gowrie. These rocks are best seen on the shore between St Monans and Pittenweem. Economically important coal-bearing cycles can only be seen in deep gorges and even better in temporary exposures created by opencast mining as at Braehead Quarry, Cardenden.

Braehead Quarry, Cardenden - open cast coal mine

Coal swamp reconstruction

Lochore Meadows Country Park - restored from coal mine to country park

Dynamic Earth

Folded rocks on the foreshore at St Monans

Most sedimentary rocks and lavas are originally laid down more or less horizontally. Dynamic earth forces may subsequently squeeze or break rocks to form folds or faults. They may also be uplifted to form mountains and plateaux, so that rocks formed on the sea bed can now be found on top of mountains as high as Everest! Volcanoes can build grand edifices. As soon as high ground is created, the bedrock is immediately subjected to the rigours of weathering - by wind, rain and frost. The product of this process - sands, mud, pebbles, or rock in solution, is then carried to the sea by streams and rivers. As sediment piles up layer upon layer, new rocks are formed on the seabed. And so the cycle begins again.

Unconformity at Whiting Ness , Arbroath

Dynamic forces are created by the movement of plates across the Earth's surface. Where a plate starts to stretch and split apart, the tension creates faults and allows molten magma to form volcanoes at surface, and the intrusion of volcanic pipes and dykes below ground. These processes can be observed in Iceland today. Such tensional forces were the underlying cause of the volcanic episodes between 350 and 300 million years ago and the intrusion of the dykes and sills of dolerite forming the Midland Valley Sill complex. These catastrophic forces also created the Ochil Fault and many other smaller dislocations across the area.

The compressional forces which buckled and broke the Carboniferous rocks were caused by plates colliding to form continental Europe about 300 million years ago. When plates collide, compressional forces squeeze the rock, as if in a vice. Rocks folded and faulted as a result of this event, are magnificently exposed near St Andrews on the Fife Coastal Path just east of the Maidens Rock sea stack and at St Monans.

Scotland was on the northern fringes of these catacylismic events. However the result was that Fife and Tayside, as part of the Midland Valley of Scotland, were to remain as land for most of the rest of geological time.

An earlier cycle of plate collision with resultant uplift and erosion of older rocks and subsequent deposition of new sediments on an ancient land surface is beautifully witnessed at Whiting Ness, Arbroath. The cliffs display an unconformity - or break in build up of sediments - between the conglomerates and sandstones of Late Devonian age and sediments of Early Devonian age. Both sets of strata were laid down by rivers but it is clearly seen that the younger deposits bank up against a quite steep land surface eroded into the Arbroath Sandstone.

When Scotland Froze

Loch Monzievaird - a kettlehole

During the last two million years, Scotland was plunged into a series of ice ages. During this period, Fife and Tayside were buried on several occasions under Greenland-like ice sheets up to one kilometre in thickness. The last great ice age lasted from about 27,000 to 13,000 years ago, although there was another cold snap 11,000 to 10,000 years ago when this area again froze but was not covered by an ice sheet.

The last great ice sheet started to form about 27,000 years ago on Rannoch Moor and in the corries of the Highlands. Ice streams overwhelmed the lower ground of Fife and Angus around Loch Leven and the Forth as well as the straths of the Eden, Earn, Tay and Strathmore. It also buried the Ochils and Sidlaws and even the mighty Highlands. The ice flowed eastwards from the Highlands across Fife and Tayside. It scoured the landscape like emery paper using the rock debris incorporated within the lower layers of ice as it advanced. Some debris remains today as `foreign' stones called erratics such as the Humlie on the shore at Cambo in east Fife. Glacial striae record the minor scale of erosion on many exposed rock surfaces.

U-shaped valley, Gleneagles

U-shaped valleys, of which Gleneagles is the best upland example, were created as the ice drove south-eastwards into the Ochil Hills. Crag-and-tail features were fashioned from harder rock such as the volcanic plug of East Lomond where the eastern lee slopes of the hill were protected from erosion. The ice left behind its ground moraine of till or boulder clay, a hard melange of stones and boulders in a matrix of clay, silt and sand which covers much of the lowland landscape. The surface of this deposit has been shaped by ice into the streamlined drumlin landforms, seen well in Stratheden and around Clackmannan, Dunfermline and Inverkeithing.

As the climate became warmer, the ice sheet receded. Eventually the climate became less arctic, and water liberated from the ice flowed copiously cutting meltwater channels beneath and at the margins of the ice. An example is Bowden Hill meltwater channel on the south side of Stratheden. These waters also laid down sands and gravels as the ice retreated.

Sediment was also deposited in tunnels in the ice and along crevasses, so when the ice melted completely, winding ridges called eskers emerged. Good examples are seen at Lindores Loch, Windygates and beside Dighty Water near Dundee. Loch Monzievaird, near Crieff marks the site where a large chunk of ice became abandoned in the outwash of sand and gravel. The ice then melted to form a hollow called a kettlehole. The original land surface was flat, but subsided when ice melted. The hollow then filled with water to form a loch.

Changing Shorelines

Kincraig raised beaches

About 16,000 years ago, the last ice sheet started to melt and retreat westwards back to the Highlands from its maximum limit offshore of Fife Ness. Then the sea-level was up to 45m higher than present day, falling to present level or below around 10,000 years ago. It rose again to as much as 10m above present about 6,500 years ago before falling again. These changes in sea level produced staircases of raised beaches and associated marine deposits, best seen in Fife at Kincraig near Elie. About 13,000 years ago the sea reached inland up the Forth as far as Aberfoyle; along the Earn as far as Crieff and up the Tay Estuary to Scone Palace, above Perth. What less than 6,500 years ago were intertidal mudflats now form the flat fertile farmland along most of the estuaries now known as the carse. The old cliffline which usually backs the carselands is a very conspicuous feature of the present-day landscape.

Modern intertidal mud and sand flats in today's estuaries can be just as extensive. Often they are flanked by sand dunes as at Tentsmuir, Barry, Lunan Bay and the unique Montrose Basin.

Along the Fife and Angus coast, relentless battering by the sea has produced some of the most interesting landforms. On rocky shores, the explosive trapping and release of air by the waves as well as mechanical erosion by current transported debris has picked out any weaknesses in the rocks, whether joints, faults or softer beds. Headlands and cliffs show a wealth of erosive features such as geos, caves, some with blowholes, coves, arches and stacks.

River Tay in flood, Perth

There is usually a wave cut platform in front of these features. Among the most famous stacks are the Rock and Spindle at St Andrews and the Deil's Heid at Arbroath. The Needle E'e, also at Arbroath and the pink Elephant Rock near the Boddin limekilns are the best arches. The Gaylet Pot at Auchmithie is a spectacular blowhole formed when the cave roof collapsed. Some features are modern but others like the Needle E'e were created when sea level was higher. However, the roughest seas at high tide still can pass through the E'e.

It is in the estuaries, where the sea and the rivers meet, that nature can show its full venom. The worst floods on the Tay at Perth usually come when rapid snow melt in the Highlands coincides with the highest tides. Such floods wash away flood defences, railway embankments and bridges, strip soil from fields, cut new braid channels as in the island at Scone Palace, and cause general disruption.

Deil's Head Stack, Arbroath

Weather and Erosion Play their Part

Landslides on Craig Rossie

Weathering and erosion have also left their mark on the landscape. During the cold snap 11,000 to 10,000 years ago, organic soils were stripped away and the rocks and subsoils opened up to freeze/thaw effects. Movement of the impressive landslides at Craig Rossie was triggered by these conditions, where the scarp face cut in lava slipped along a series of rotational failure surfaces. At other locations, notably on the Lomond Hills, rock has been broken up by weathering and now forms angular blocks in a scree slope.

One of the most graceful landmarks in the Lomond Hills is the rock column known as Carlin (witch) Maggie below Bishop Hill. On the escarpment of the sill, this stack has been left isolated by the toppling of the adjacent jointed columns of dolerite. A popular local rock-climb, she has already lost her tourie (pom-pom) off the top. When gravity will destroy the rest we can only speculate.

Carlin Maggie

Changing Landscapes

Collace Quarry at Dunsinane Hill

Man has altered the landscape in many ways. The most significant changes have resulted from the general effects of deforestation, intensive farming and mineral extraction. Wetlands and peatlands have been also been lost or damaged, as in the commercial peat cutting at Moss Morran.

Rock quarries, opencast coal mines and sand and gravel pits have all altered the landscape to some degree. Many of these voids have been landfilled, so bringing the land back into profitable use. Like the many worked-out collieries, the once

New hill at Lochore created with spoil from Westfield opencast site

Kingoodie Quarry

240m deep Westfield opencast, has disposed of its excess spoil (redd) to bings, thus creating new landscape features. Some of Westfield's redd now forms the 2km long `Three Sevens Hill' between Lochore and Auchterderran. Another form of restoration is demonstrated at Kingoodie Quarry where nature has reclaimed the site. This disused quarry is now managed as a nature reserve.

Recent proposals to extend Collace Quarry towards Macbeth's castle on Dunsinane Hill were fought on grounds of damage to the landscape and our cultural heritage. It is also possible that the mythical Stone of Destiny hidden by the monks from King Henry 1st, still lies in some nook or cranny on Dunsinane. If the hillside were to be removed by quarrying, the whereabouts of the Stone would remain a mystery for ever.

Scottish Natural Heritage and the British Geological Survey

Scottish Natural Heritage is a government body. Its aim is to help people enjoy Scotland's natural heritage responsibly, understand it more fully and use it wisely so that it can be sustained for future generations.

Scottish Natural Heritage
12 Hope Terrace
Edinburgh EH9 2AS

The British Geological Survey maintains up-to-date knowledge of the geology of the UK and its continental shelf. It carries out surveys and geological research.
The Scottish Office of BGS is sited in Edinburgh. The office runs an advisory and information service, a geological library and a well-stocked geological bookshop.

British Geological Survey
Murchison House,
West Mains Road
Edinburgh EH9 3LA

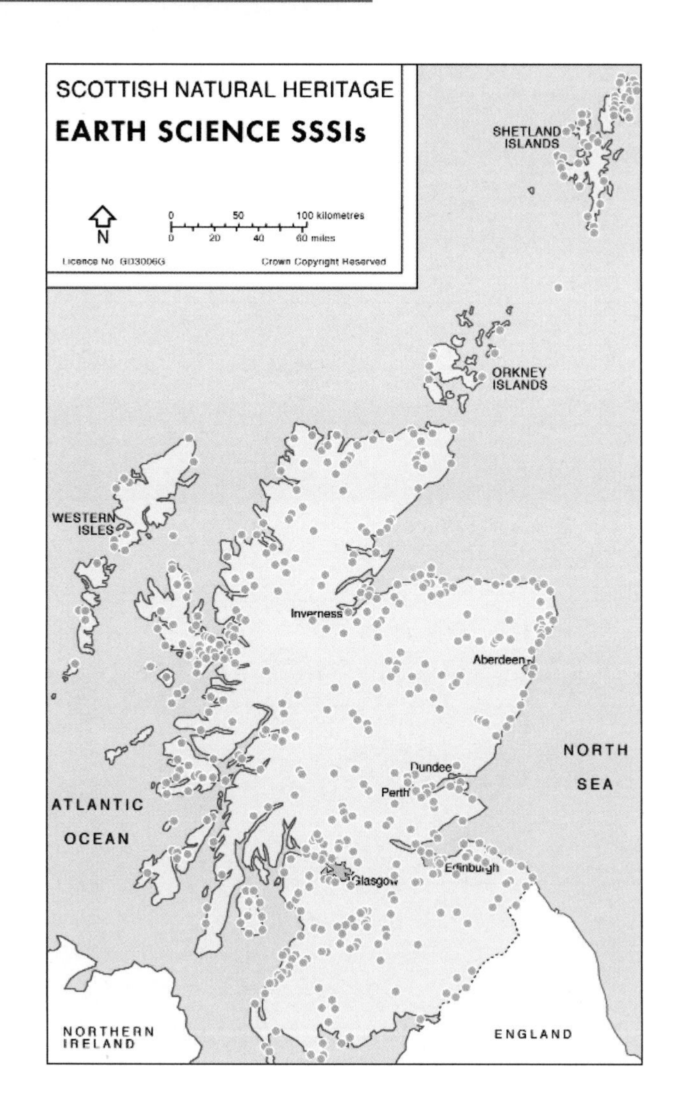

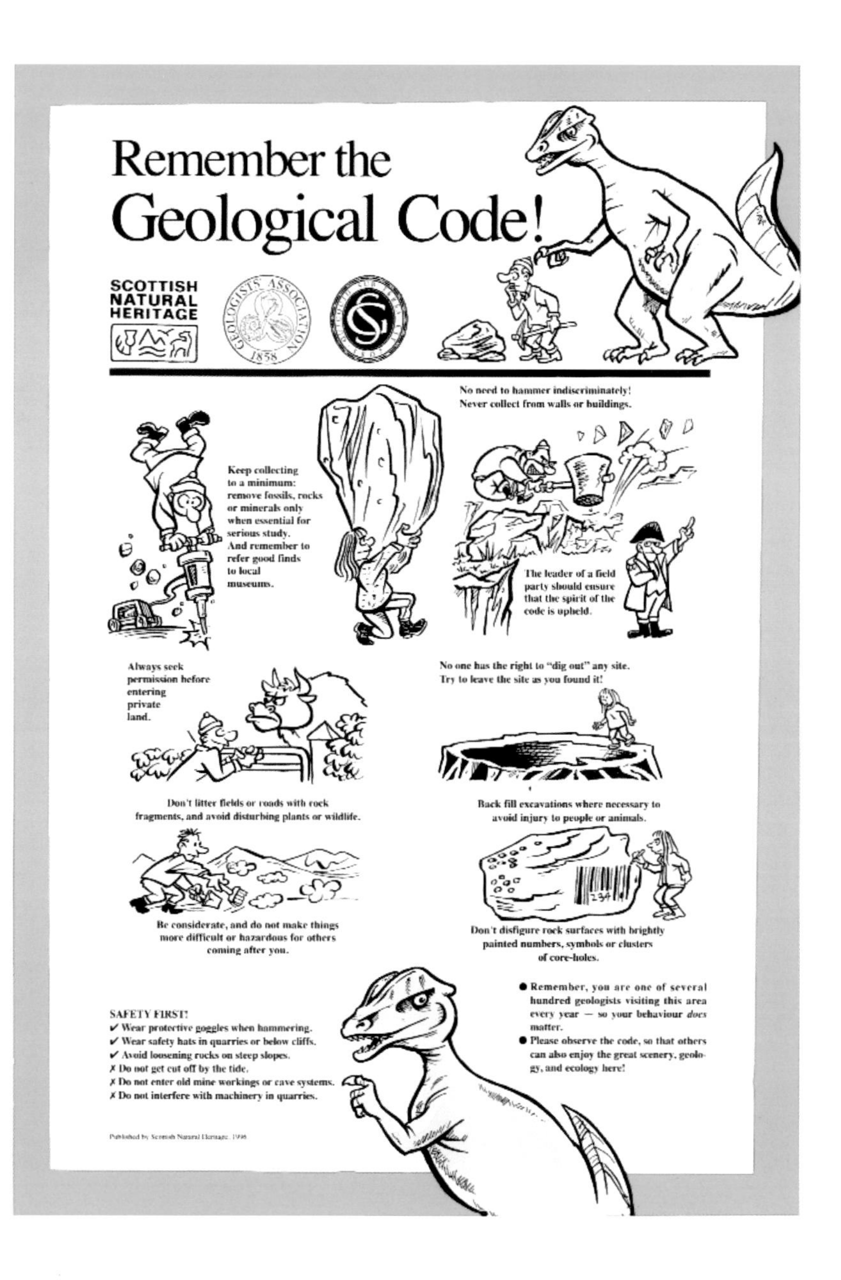

Remember the
Geological Code!
SCOTTISH
NATURAL
HERITAGE
GEOLOGISTS' ASSOCIATION
1858
No need to hammer indiscriminately!
Never collect from walls or buildings.
Keep collecting
to a minimum:
remove fossils, rocks
or minerals only
when essential for
serious study.
And remember to
refer good finds
to local
museums.
The leader of a field
party should ensure
that the spirit of the
code is upheld.
Always seek
permission before
entering
private
land.
No one has the right to "dig out" any site.
Try to leave the site as you found it!
Don't litter fields or roads with rock
fragments, and avoid disturbing plants or wildlife.
Back fill excavations where necessary to
avoid injury to people or animals.
Be considerate, and do not make things
more difficult or hazardous for others
coming after you.
Don't disfigure rock surfaces with brightly
painted numbers, symbols or clusters
of core-holes.
● Remember, you are one of several
hundred geologists visiting this area
every year — so your behaviour *does*
matter.
● Please observe the code, so that others
can also enjoy the great scenery, geolo-
gy, and ecology here!
SAFETY FIRST!
✔ Wear protective goggles when hammering.
✔ Wear safety hats in quarries or below cliffs.
✔ Avoid loosening rocks on steep slopes.
✗ Do not get cut off by the tide.
✗ Do not enter old mine workings or cave systems.
✗ Do not interfere with machinery in quarries.
Published by Scottish Natural Heritage, 1996

Also in the Landscape Fashioned by Geology series...

If you have enjoyed Fife & Tayside why not find out more about the geology of some of Scotland's distinctive areas in our Landscape Fashioned by Geology series. Each book helps you to explore what lies beneath the soils, trees and heather with clear explanations, stunning photographs and illustrations. The series, which is produced in collaboration with the British Geological Survey, is written by experts in a style which is accessible to all.

Arran and the Clyde Islands

The diverse landscapes of Arran and the Clyde Islands mark the boundary between Highland and Lowland. Discover the ancient secrets and the appeal of these well-loved islands.
David McAdam & Steve Robertson
ISBN 1 85397 287 8 pbk 24pp £3.00

Cairngorms

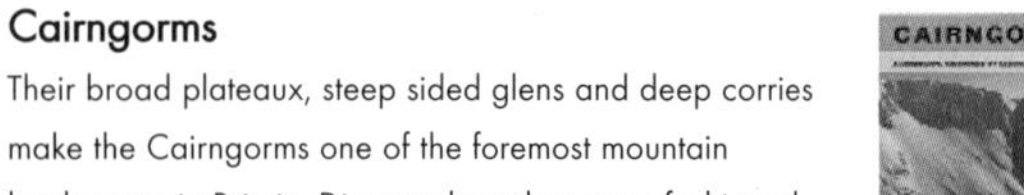

Their broad plateaux, steep sided glens and deep corries make the Cairngorms one of the foremost mountain landscapes in Britain. Discover how they were fashioned by weathering, glaciers and rivers.
John Gordon, Vanessa Brazier,
Rob Threadgold & Sarah Keast
ISBN 1 85397 086 7 pbk 28pp £2.00

East Lothian and the Borders

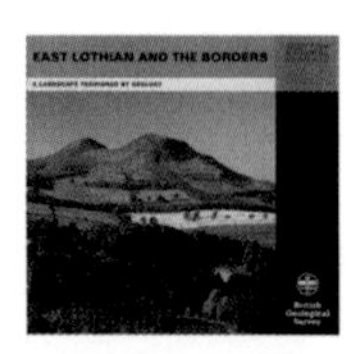

Underneath the calm facade of south east Scotland's fertile plains and rolling hills lies a complex structure, which reflects an eventful geological history.
David McAdam & Phil Stone
ISBN 1 85397 242 8 pbk 26pp £3.00

Edinburgh

Some of the most important discoveries in geological science were made in and around Edinburgh. This booklet is full of startling facts about the Capital's geological past.
David McAdam
ISBN 1 85397 024 7 pbk 28pp £2.50

Loch Lomond to Stirling

The heart of Scotland encompasses some of the most diverse landscapes in Scotland. From the low Carse to the mountain tops - find out how these modern landscapes reflect the geological changes of the past.
Mike Browne & John Mendum
ISBN 1 85397 119 7 pbk 26pp £2.00

Orkney and Shetland

These northern outposts of Scotland hold a great fascination for the geologist. Starting 3 billion years ago, their story tells of colliding continents, bizarre lifeforms and a landscape which continues to be eroded by the pounding force of the Atlantic.
Clive Auton, Terry Fletcher & David Gould
ISBN 1 85397 220 7 pbk 24pp £2.50

Skye

Skye is one of Scotland's most popular tourist destinations, and deservedly so. But what would Skye be without the jagged peaks of the Cuillins or the intriguing rock formations of the Quirang? In many ways it is the geology of Skye that attracts it's visitors and this booklet helps you to understand how the mountains, rocks and lochs were formed.
David Stephenson & Jon Merritt
ISBN 1 85397 026 3 pbk 24pp £2.50

Scotland: the creation of its natural landscape

Scotland: the Creation of its Natural Landscape provides a wealth of information on how Scotland was created and the events that took place there through the aeons. But the story doesn't stop back in the mists of time, it continually unfolds and this book provides up to the minute information on geological events taking place beneath our feet, It also provides a history of geological science and highlights the enormous contribution Scots geologists have made to the world.
Alan McKirdy and Roger Crofts
ISBN 1 85397 004 2 pbk 64pp £7.50

Series Editor: Alan McKirdy (SNH)

Other books soon to be produced in the series include:

Mull and Iona	North West Highlands
Rum	Parallel Roads of Glen Roy

SNH Publication Order Form

Title	Price	Quantity
Arran & the Clyde Islands	£3.00	
Cairngorms	£2.00	
East Lothian & the Borders	£3.00	
Edinburgh	£2.50	
Loch Lomond to Stirling	£2.00	
Orkney & Shetland	£2.50	
Skye	£3.95	
Scotland: the Creation of its natural landscape	£7.50	

Postage and packaging: free of charge within the UK.

A standard charge of £2.95 will be applied to all orders from the EU.

Elsewhere a standard charge of £5.50 will apply.

Please complete in **BLOCK CAPITALS**

Name ______________________

Address ______________________

____

____

Post Code ______________________

Type of Credit Card VISA ☐ EUROCARD MasterCard ☐

Name of card holder ______________________

Card Number ☐☐☐☐ ☐☐☐☐ ☐☐☐☐ ☐☐☐☐

Expiry Date ☐☐ ☐☐

Send order and cheque made payable to Scottish Natural Heritage to:

Scottish Natural Heritage, Design and Publications, Battleby, Redgorton, Perth PH1 3EW

E-mail: pubs@redgore.demon.co.uk www.snh.org.uk

Please add my name to the mailing list for the SNH Magazine ☐

Publications Catalogue ☐